DISASTER DIARIES

Surviving the VOLCANO

HEAR MY STORY

Linda Barghoorn

Crabtree Publishing
crabtreebooks.com

DISASTER DIARIES

Author: Linda Barghoorn

Editorial director: Kathy Middleton

Editors: Sarah Eason, Jennifer Sanderson, and Ellen Rodger

Proofreaders: Tracey Kelly, Melissa Boyce

Editorial director: Kathy Middleton

Design: Paul Myerscough

Cover design: Margaret Amy Salter

Photo research: Rachel Blount

Production coordinator and Prepress technician: Tammy McGarr

Print coordinator: Katherine Berti

Consultant: John Farndon

Produced for Crabtree Publishing Company by Calcium

Photo Credits:
t=Top, c=Center, b=Bottom, l= Left, r=Right

Inside: Shutterstock: Zahirul Alwan: p. 8; Desislava Antonova: pp. 12–13; Artsuedan: p. 23t; Francesco Cepolina: p. 25; ChameleonsEye: p. 27; CRStocker: p. 10; Henner Damke: p. 26; Daxiao Productions: p. 21; Designua: pp. 14–15t; Fboudrias: p. 4; Fotos593: p. 7r; GeoJGomez: p. 17r; Hanohiki: pp. 6–7; J. Helgason: p. 5; Ammit Jack: p. 13t; Jess Kraft: pp. 28–29; Linnas: p. 20; MNStudio: p. 19; Andres Navia Paz: p. 29t; Valeriy Poltorak: pp. 1, 18; RosieYoung: p. 15b; Sailorr: p. 9; Smlopezz: pp. 22–23; Zef Art: pp. 16–17; Wikimedia Commons: Tom Casadevall, USGS: p. 11; USGS: p. 24.

Cover: Shutterstock

Publisher's Note: The story presented in this book is a fictional account based on extensive research of real-life accounts, with the aim of reflecting the true experience of victims of natural disasters.

Library and Archives Canada Cataloguing in Publication

Title: Surviving the volcano : hear my story / Linda Barghoorn.
Names: Barghoorn, Linda, author.
Description: Series statement: Disaster diaries | Includes index.
Identifiers: Canadiana (print) 20200151363 | Canadiana (ebook) 20200151371 | ISBN 9780778769965 (hardcover) | ISBN 9780778771203 (softcover) | ISBN 9781427124494 (HTML)
Subjects: LCSH: Ruiz, Nevado del (Colombia)—Eruption, 1985—Juvenile literature. | LCSH: Volcanoes—Colombia—Juvenile literature. | LCSH: Volcanoes—Juvenile literature. | LCSH: Disaster victims—Juvenile literature.
Classification: LCC QE523.R85 B37 2020 | DDC j551.2109861/3—dc23

Library of Congress Cataloging-in-Publication Data

Names: Barghoorn, Linda, author.
Title: Surviving the volcano : hear my story / Linda Barghoorn.
Description: New York, New York : Crabtree Publishing Company, [2020] | Series: Disaster diaries | Includes bibliographical references and index.
Identifiers: LCCN 2019057535 (print) | LCCN 2019057536 (ebook) | ISBN 9780778769965 (hardcover) | ISBN 9780778771203 (paperback) ISBN 9781427124494 (ebook)
Subjects: LCSH: Volcanoes--Colombia--Juvenile literature. | Volcanic eruptions--Juvenile literature. | Survival--Juvenile literature. | Ruiz, Nevado del (Colombia)--Eruption, 1985--Juvenile literature. | Armero (Colombia)--History--20th century--Juvenile literature.
Classification: LCC QE524.2.C7 B37 2020 (print) | LCC QE524.2.C7 (ebook) | DDC 551.2109861--dc23
LC record available at https://lccn.loc.gov/2019057535
LC ebook record available at https://lccn.loc.gov/2019057536

Crabtree Publishing

crabtreebooks.com 800-387-7650

 In Canada: We acknowledge the financial support of the Government of Canada through the Canada Book Fund for our publishing activities.

Published in Canada
Crabtree Publishing
616 Welland Avenue
St. Catharines, Ontario
L2M 5V6

Published in the United States
Crabtree Publishing
347 Fifth Avenue
Suite 1402-145
New York, NY 10016

Hardcover 978-0-7787-6996-5
Paperback 978-0-7787-7120-3
Ebook (pdf) 978-1-4271-2449-4

Printed in Canada/042024/CPC20240426

Contents

Volcanoes and Their Victims

Volcanoes are majestic and mighty, but they can also be incredibly dangerous and destructive. When a volcano **erupts**, it can gush red-hot liquid rock and fling out glowing stones and a choking cloud of scorching ash. Volcanoes can create impressive spectacles of fiery **lava** fountains or volcanic lightning. They can also trigger deadly **mudslides**, **landslides**, **avalanches**, and **tsunamis**, and can create ash clouds that can threaten human populations.

The Life Span of a Volcano

Volcanoes can last many thousands of years. They are considered to be active, dormant, or extinct, depending on when they last erupted. Of the three types of volcanoes, active volcanoes pose the largest threat. These volcanoes have erupted within recent centuries and are likely to erupt again.

Named after Vulcan, the Roman god of fire, volcanoes literally help Earth let off steam.

Dormant volcanoes are still active deep below Earth's surface. Although they have not erupted in many years, they could do so again. Extinct volcanoes have not erupted for 10,000 years and are not expected to erupt again.

The fragments in a volcanic ash cloud can travel thousands of miles, disrupting the lives of people in many countries.

Around the World

Earth is home to around 500 active volcanoes. At least 50 volcanoes erupt each year. Some eruptions are so weak that they are barely noticeable. Others happen where few people live. However, some volcanoes can erupt with such power that they change landscapes, damage cities and towns, and threaten people both nearby and farther away. Ash and mud can poison water supplies. The mudslides and landslides triggered by eruptions can destroy bridges, roads, and forests.

LUIS'S STORY

In this book, you can find out what it is like to live through a **natural disaster** by reading the **fictional** story of Luis, a young boy caught up in the volcanic eruption of Nevado del Ruiz in Colombia. Look for his story on pages 6–7, 12–13, 16–17, 22–23, and 28–29.

LUIS'S STORY:
Before the Volcano

My name is Luis, and I am Colombian. I used to live in a small town called Armero. It was tucked into a beautiful valley in an area of land made of rich volcanic soil. Like many others around us, my family had lived there for generations. Most families were large, and everyone was expected to help out on the farms after school and on weekends. When I was not doing homework or farm chores, I liked to play soccer, which we call football, with my friends Alessandro and Mateo. My favorite team is from Bogotá—they are called the Millonarios.

The nearby Nevado del Ruiz volcano has always been known as "the Sleeping Lion." I guess this is because it spends much of its time not doing very much. But every once in awhile it wakes up and erupts with a loud roar. Our history books told us it had erupted twice.

Soccer is a national pastime in Colombia. We played soccer without realizing that Ruiz was about to erupt.

The last time Ruiz erupted was more than 170 years ago. Some people were killed by the mudflows that swept through the valleys. Nothing had happened since, and Armero was even built on the **deposits** of mud from that eruption. No one seemed to spend much time thinking about our volcano.

When I was 12 years old, Ruiz began to rumble again. It rumbled for months, but we were told that there was nothing to worry about. One day, several climbers who had reached the top of the **crater** reported that they saw small clouds of gases coming out of the volcano. A month later, three small **tremors** shook the ground slightly. My parents were nervous, but the government was not convinced that Ruiz was a threat to anyone. After all, it was more than 40 miles (64 km) away from our town. We had no idea that the Sleeping Lion was stirring to life again.

I feel sad when I think about our lives before the volcano erupted. I don't think I'll ever feel safe again.

Many farmers in the region made their living growing cotton on the ***fertile*** *farmland.*

Trails of Destruction

Volcanic ash covered the city of Yogyakarta in Indonesia when Mount Kelud erupted in 2014.

Throughout human history, volcanic eruptions have created epic disasters. Many thousands of people have been killed. Damage to buildings, roads, bridges, and crops has cost billions of dollars. Sometimes, there may be signs that a volcano is about to erupt, while at other times, volcanoes erupt without warning. Their power can be swift and devastating.

Ash Clouds

One of the most dangerous consequences of an eruption are the thick clouds of ash that are pushed into the **atmosphere**. These ash clouds contain billions of particles of rock and glass. They make it impossible to breathe as they are inhaled, or breathed in. They can bury cities and towns and damage machinery. When they get into airplane engines, they can cause plane crashes.

Ash clouds can block out the Sun's rays, reducing temperatures around the world. When this happens, frost destroys crops, and the cold can keep crops from ripening. The gases in the ash can mix with water to create **acid rain**. This damages crops and kills livestock.

VICTIMS OF ROCK AND ASH

More than 2,000 years ago, Mount Vesuvius erupted in ancient Rome. The city of Pompeii, which lay at the foot of the volcano, was destroyed within hours. Millions of tons of ash buried the city. Everyone in the city was killed.

In 1748, explorers unearthed the ancient city of Pompeii. They found it much as it had been just moments before the eruption.

Magma, Mud, and Landslides

Magma is hot, liquid rock. It is pushed to Earth's surface during a volcanic eruption. When magma erupts, it is called lava. Lava flows down a volcano in a dangerous river of liquid rock, and as it flows away, it can cool and turn into solid rock. Although lava flows are often slow enough to allow people time to escape, they can flatten forests, block rivers, and destroy bridges and roads. Volcanic eruptions give off heat, which can cause snow or glaciers on a volcano's peak to melt. This creates a fast-moving mixture of water, rock, and lava called a **lahar**. Sometimes a volcano's side collapses during an eruption, causing a deadly landslide or mudslide.

How Volcanoes Work

Earth is made of three main layers: the core, the mantle, and the crust. The crust is Earth's outer layer, and ranges from 3 miles (5 km) to 44 miles (70 km) thick. Compared to the rest of the planet, this crust is very thin—like the outer skin of an apple.

Earth's Mantle

Earth's largest layer, the mantle, lies directly under the outer crust. It is made up mostly of solid rock, but some of the rock is so hot that it melts. This liquid rock is the magma that erupts. Pressure from deep within the planet pushes this magma toward Earth's surface.

After an extremely powerful eruption, **toxic** combinations of rock, ash, and gas flow out of the volcano. This is known as a pyroclastic flow. Pyroclastic flows are fiery hot and can move at tremendous speed down the sides of volcanoes, burning everything in their path.

When magma finds a weak spot in Earth's crust, it explodes to the surface in a volcanic eruption.

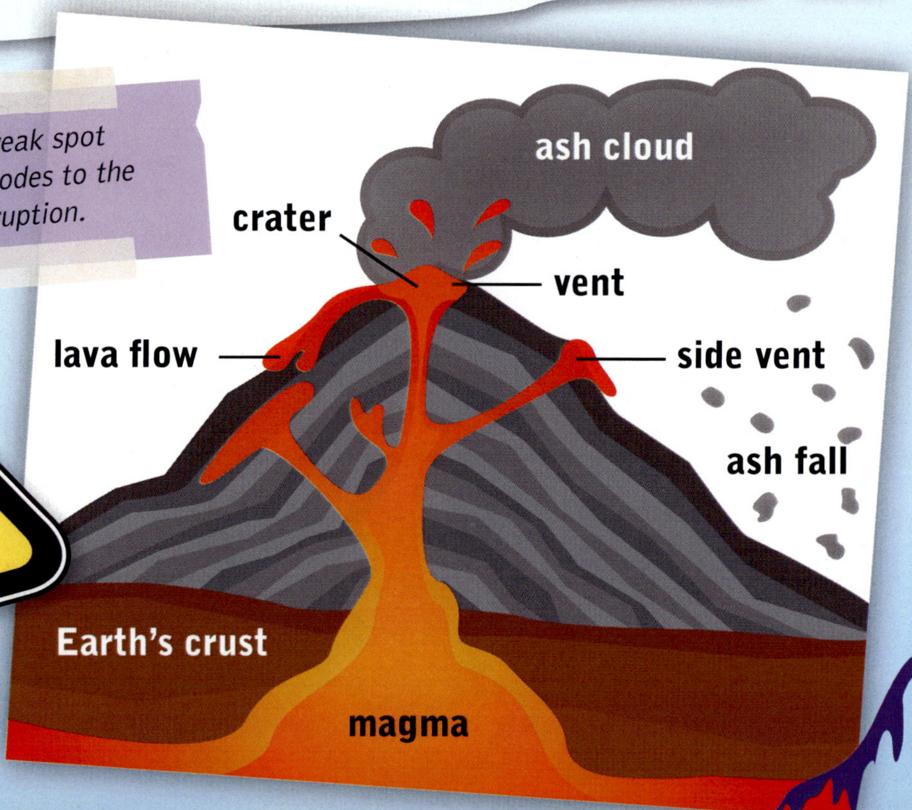

Lahars—made up of pyroclastic material, rocky **debris**, and water—flow down a volcano's slopes in a thick sludge that resembles cement.

Types of Volcanoes

When volcanoes erupt, they leave behind different structures. **Cinder** cones are circular or oval in shape and made of cinders. They form quickly but are usually very small. Composite volcanoes are formed of mixtures of ash, cinder, and lava. They also are cone-shaped but can grow after each eruption to be huge in size. Shield volcanoes are vast, shallow shield shapes formed by runny lava that spreads fast. Lava domes are like shield volcanoes but made from sticky lava that piles up in a dome shape.

VICTIMS OF THE LANDSLIDE

In 1980, an earthquake took place below Mount St. Helens in Washington. The earthquake triggered a volcanic eruption that tore off the top of the volcano, causing the largest landslide in recorded history. It sent an enormous cloud of gas and rock down the mountainside. Forests were flattened, and 185 miles (298 km) of road and 200 homes were destroyed.

LUIS'S STORY: The Volcano Erupts

After almost a year of small earthquakes and explosions of steam coming from its crater, Nevado del Ruiz finally violently exploded one afternoon. I was in school, and we all heard the loud blast far off in the distance. We rushed to the windows to look outside but couldn't see anything unusual. We were excited and nervous about what had happened, but there were no alarms. School continued for another hour before classes ended for the day. I met my brother Nicolas and my sister Maria to walk the short distance home.

We were in the kitchen eating a snack when pieces of rock and ash began raining down on our house. Papa was still in the fields working. When Mama turned on the radio to listen to the news, the mayor was making an announcement. He said that everyone should remain calm. He advised people to stay in their homes to avoid the hot, falling ash that could injure them.

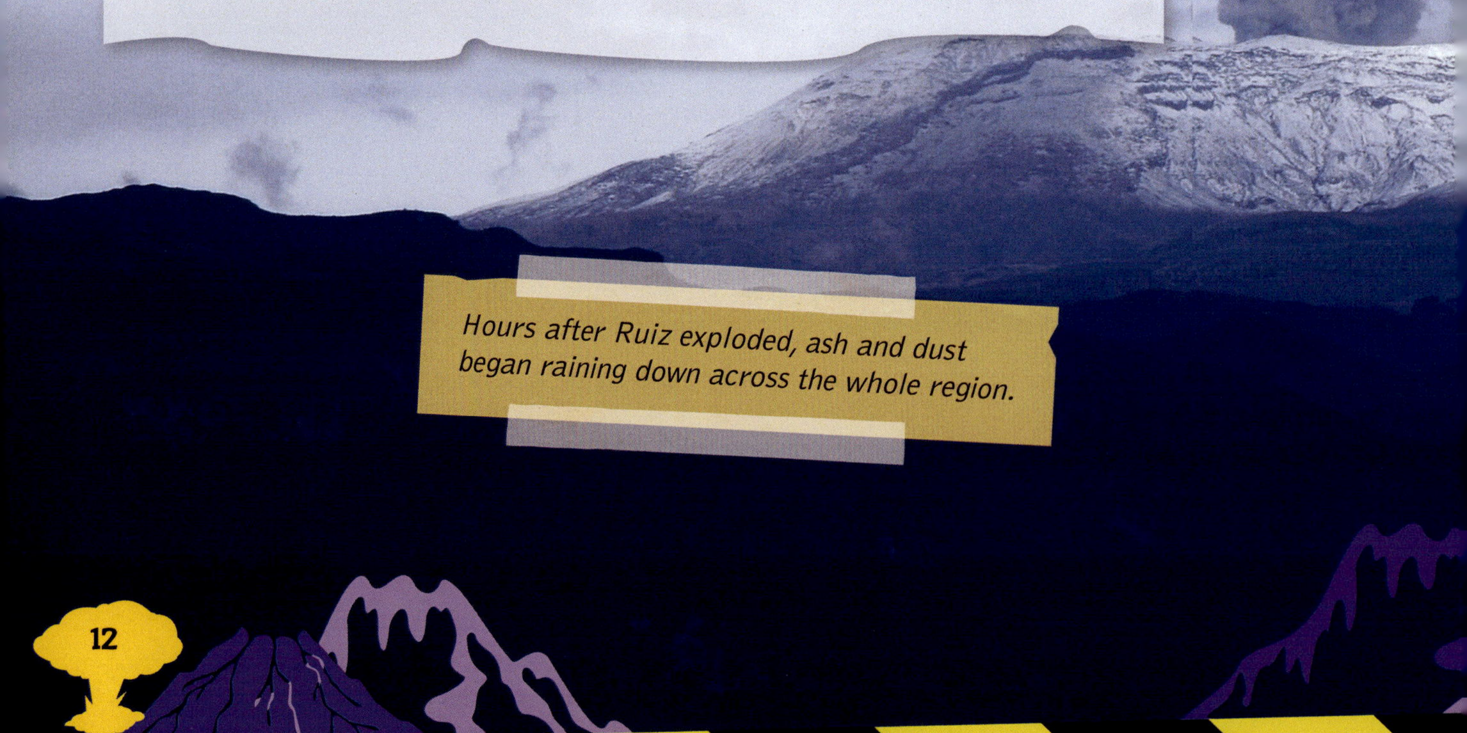

Hours after Ruiz exploded, ash and dust began raining down across the whole region.

When Papa arrived home, his shirt was full of small holes from where the ash had landed on him. His whole body was covered with a fine layer of ashy dust. His hair looked white, like an old man's. Mama cooked dinner while we listened for updates on the radio. A public loudspeaker that was nearby crackled to life as our priest made an announcement. He, too, wanted to reassure us all that there was no need to panic.

A fine blanket of ash covered everything in town, from car windshields to our houses.

Soon, the ground and trees outside were coated with a fine layer of volcanic ash. We even joked that it looked like snow. As we got ready to sit down to dinner, the Red Cross organization made an urgent radio announcement, ordering the town to evacuate. Mama and Papa grabbed a few things and we got ready to leave. But within minutes, the ash stopped falling, and the evacuation was called off. Although my parents were worried, they decided to follow the advice, and we nervously settled in for the evening. Finally, I went to sleep, hoping everything would be back to normal in the morning.

The Causes of Volcanoes

Earth's crust is made up of massive slabs of land called **tectonic plates**. These plates are always moving, and as they move, they spread, collide, or slide past each other. This happens so slowly that people do not feel anything.

Where Volcanoes Occur

Most volcanoes occur at the boundaries between these tectonic plates. However, some volcanoes occur in the interior of plates at an area called a **hot spot**.

Under Pressure

When a volcano erupts, the material that it emits, or gives off, can include magma, ash, cinder, rock debris, and gas. The pressure of the gases that are dissolved in the magma is usually what drives the eruption. How explosive an eruption is depends on whether the magma inside the volcano is runny or sticky when pressure builds up on it before an eruption.

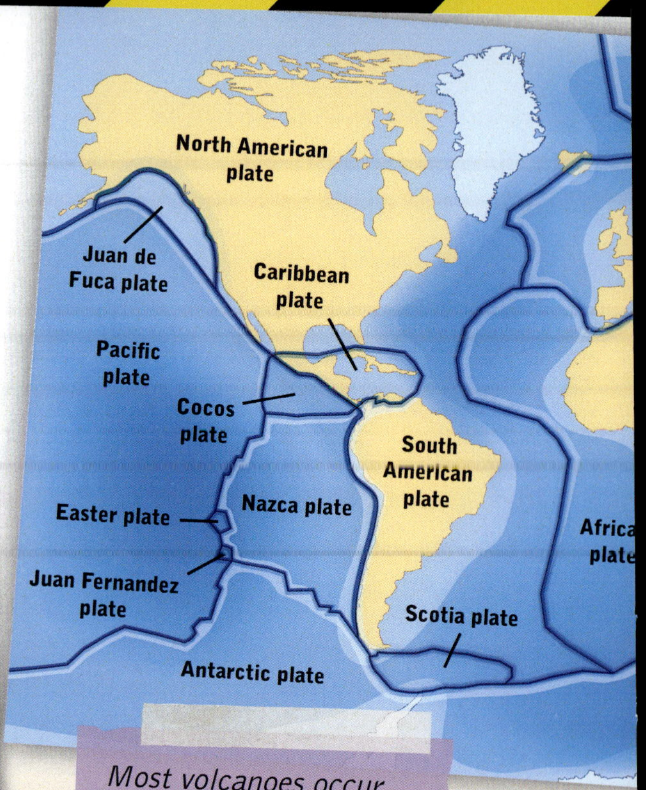

Most volcanoes occur close to the cracks dividing tectonic plates.

VICTIMS OF A TSUNAMI

One of the most violent volcanic eruptions in recent history took place in Indonesia in 1883. It was four times more powerful than the largest bomb humankind has ever exploded. It was so powerful that it destroyed the island it sat on. It triggered a series of tsunamis that killed 36,000 people and devastated huge areas of coastal landscapes and towns.

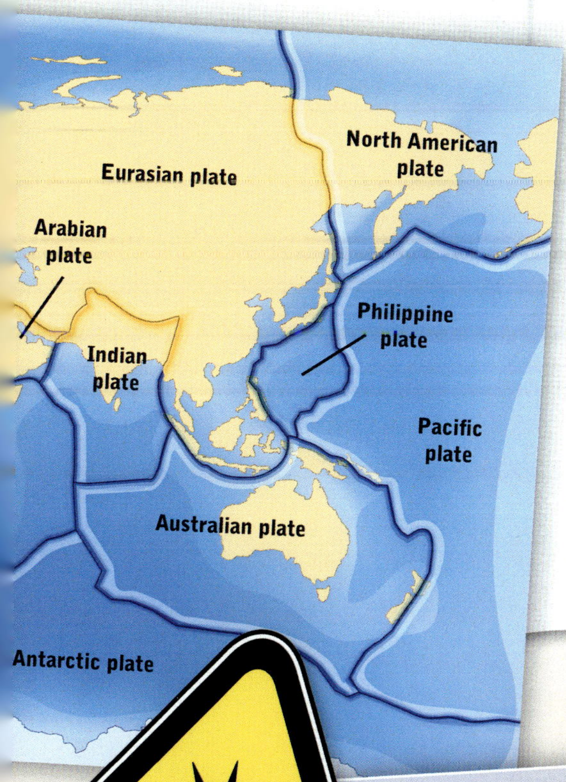

Volcanic Gases

Water vapor and carbon dioxide are given off during an eruption in a large cloud known as a gas **plume**. Sulfur dioxide is also sometimes present. This is an invisible, foul-smelling gas. Gas plumes are blown about by winds that can help them circle Earth in a week. Our planet's entire atmosphere was created by this process of outgassing—the release of gases from deep within our planet. The air we breathe began as volcanic gas.

When volcanoes erupt, they emit smoke and steam that is full of sulfur.

LUIS 'S STORY:
Fleeing for Our Lives

I was in a deep sleep when I felt someone shaking me, trying to wake me up. It was Papa, who told me to get dressed quickly because we had to leave right away. He had been listening to the radio while we slept and heard news that other towns were being evacuated. He had decided it was time for us to leave. He didn't want to wait any longer.

He handed me my clothes and shoes, which I rushed to put on. I could hear Mama in the next room with my sister. She was trying to get her dressed while Maria sobbed, afraid. There were no lights on, and the radio was silent. The electricity had gone out. We hurried around in the darkness as Papa used a flashlight to help guide us. From somewhere far away, I could hear a rumbling noise like the sound of low, distant thunder.

When volcanoes erupt at night, people are often asleep. Even if they wake up due to the sound of explosions, they may still be drowsy. This means they have less chance to react quickly and escape.

The Moon was out, but the village was in darkness as most people slept. As my eyes adjusted to the dark, I could see a few others moving. Where were Alessandro and Mateo and their families, I wondered? As the rumbling grew louder, Papa directed us toward the nearest hill, away from the village. We climbed and ran, stumbling as we tried to make our way. "The mud is coming," Papa said, "We need to hurry."

As we clambered up the hill, the rumbling sound grew louder and louder. Papa had to shout as he tried to guide us up the hill. The ground began to shake, and I could hear loud crashes close by. As we made it to the top of the hill, we turned to look at the village below us. In the moonlight, I could see something moving in the distance, coming closer and closer. It was a gigantic wall of mud roaring down the valley. We huddled together, frightened and alone on the hilltop, as the mud approached our town.

Mudslides can race down hillsides and smother any towns and villages below them in deep, deadly mud.

Where Volcanoes Happen

Most volcanoes are found along a belt called the Ring of Fire, which circles the Pacific Ocean. The Ring of Fire runs north from the base of Chile up to Alaska, then south below Siberia, and on to Japan, the Philippines, and then to New Zealand. It is where the huge Pacific plate, which lies beneath the ocean, joins several other tectonic plates that lie beneath Earth's continents.

The Ring of Fire

The Ring of Fire represents the largest risk of volcanic activity for people in many countries. The countries at risk range from Chile to Japan and many nations of the Pacific Islands, such as the Solomon Islands. A series of active volcanoes stretches along the west coast of the United States, from Alaska to Oregon and California. Many of the world's most devastating volcanic events have occurred along the Ring of Fire, including Mount Tambora and Krakatoa in Indonesia; Novarupta in Alaska and Mount St. Helens in Washington; Mount Ruiz in Colombia; and Mount Pinatubo in the Philippines.

Mount St. Helens is an active volcano in the Ring of Fire. Its last eruption was in 2008.

Mauna Loa has been erupting for more than 700,000 years.

Mighty Mauna Loa

The Hawaiian Islands are a chain of volcanoes that formed over a hot spot near the center of the Pacific plate. Some of these volcanoes have emerged as land, but others stay submerged, or underwater. The chain of volcanoes formed because the Pacific plate moves over the top of the hot spot. As the plate moved over the hot spot, it burned through the plate and caused volcanoes to form. Mauna Loa, in Hawaii, began as an underwater volcano that slowly emerged from the sea over a period of hundreds of thousands of years. Measured from its base on the ocean floor to its peak, it is taller than Mount Everest. It is the world's largest active volcano.

VICTIMS OF AN ASH CLOUD

The Mount Tambora eruption in 1815 was the most powerful in recorded history. Its ash cloud soared 25 miles (40 km) into the atmosphere before being blown across the globe. As sunlight was blocked from reaching Earth, global temperatures dropped more than 5 degrees Fahrenheit (3 degrees Celsius). This caused crops to fail in many countries. Global food shortages meant that thousands of people died of starvation.

Volcano Season

Recent scientific studies suggest that there could be a link between volcanic activity and Earth's **orbit** around the Sun. Volcanoes seem to be more active when ocean **tides** are at their lowest point. Since the tides are a direct effect of the daily variations in **gravitational** force as the Moon circles Earth, scientists believe there could be a connection. But much of the science behind the timing of volcanic eruptions remains a mystery.

Volcanologists are scientists who study volcanoes. These volcanologists are examining rocks near a volcano to learn more about it.

Monitoring Activity

Scientists have been developing tools to predict when a volcano is likely to erupt. They use forecasts that examine volcanic activity, including small earth tremors, the movement of magma, and the release of gases. The forecasts provide the best signals of a possible eruption. Understanding the past behaviors of a volcano can also provide valuable clues about when it will next erupt. However, volcanoes are not always predictable. The best forecasts combine regular observation of a volcano's current activity and an in-depth understanding of its past behavior.

Volcanic Explosivity Index

The Volcanic Explosivity Index—or VEI—measures the size of volcanic eruptions. It measures the height of the ash cloud, the amount of **tephra** ejected, and the strength of the blast to determine the VEI. Nonexplosive eruptions are measured at 0, the lowest end of the VEI. Super eruptions are given a VEI rating of 8, at the highest end of the index. Volcanic eruptions that are rated at VEI 7 or higher are rare. They happen about once or twice in 1,000 years. The last VEI 7 eruption was the Mount Tambora eruption, which took place in 1815. It killed 92,000 people.

VICTIMS OF THE PYROCLASTIC FLOW

Mount Pinatubo in the Philippines erupted in 1991, triggering a deadly, high-speed lahar. More than 9 cubic miles (38 cubic km) of fiery-hot ash, gas, and mud combined to form volcanic deposits up to 660 feet (200 m) thick in the surrounding lowlands. More than 200,000 people had to be evacuated to safety.

When Krakatoa erupted on March 31, 2014, it registered a VEI of 1 and caused little damage.

LUIS'S STORY: Rescue!

I don't remember falling asleep, but when I woke up the next morning I could not believe what I saw. Everything in the valley was buried under a thick blanket of mud. Trees were uprooted and lay in tangled piles. Cars and tractors had been tossed around like toys and lay twisted in the mud. Telephone and electricity lines lay scattered everywhere. Houses had been gobbled up by the mud.

No one else was around. It felt like we were alone in the world. Where had everyone gone? We were afraid to go down to the village in case more mud came. So with no way to contact anyone, we waited on the hilltop—hungry, thirsty, tired, and scared.

Huge areas of land can be covered in lava and other volcanic material after an eruption. Highways can be cut off, making it impossible for people to travel from one place to another. This can make it very difficult to get supplies to people who need them.

Early in the afternoon, a helicopter appeared in the sky. We jumped up, waving and yelling, hoping those on board would see us. They did! A rescue team pulled us up to safety. They took us to a shelter where emergency tents had been set up. A medical team rushed around taking care of people with burns on their bodies or broken limbs. We felt so lucky to be okay.

Often, the only way to get supplies to people after a volcanic disaster is to fly them in by airplane or helicopter.

In the days that followed, we learned that the whole town had been destroyed. Many people had been killed. Papa was desperate to get news of our family and neighbors. A man from our village arrived at the shelter, but he had also been separated from his family and did not know if they were alive. How could this happen? I wondered. Why didn't anyone warn us?

We spent several weeks at the shelter before being moved to another camp. Many people had started to place notices on buildings to ask for help finding their relatives. Papa did the same, posting a notice with the names of our missing family members and telling them where they could find us. The local radio station also provided information for lost relatives, so Papa added our names to the station's daily announcements.

How Science Can Fight Volcanoes

People have lived side by side with volcanoes for thousands of years. As the world's population expands, more people are likely to live near volcanoes, putting them at ever greater risk. So it is important to understand volcanic behavior in order to protect people in high-risk areas.

This volcanologist is taking samples of the molten lava of Kilauea volcano in Hawaii.

VICTIMS OF DISTANCE

Scientists have discovered that one-tenth of the world's population lives near an area of volcanic danger and that more than 800 million people live within 62 miles (100 km) of active volcanoes. Records show that between 1500 and 2017, more than 278,000 people were killed by volcanoes—that's about 540 people every year since 1500. If people are killed within 3 miles (5 km) of volcanoes, it is usually because of **volcanic bombs**. If they are killed within 3 to 9 miles (5 km to 15 km), the cause is often avalanches of hot rock, ash, and gas. Farther than 9 miles (15 km) from a volcano, people are usually killed by mudflows, tephra, or tsunamis.

Studying the gas that escapes from vents in the sides of volcanoes can tell scientists more about when they might erupt.

Volcanologists at Work

As scientific tools have become more advanced, volcanologists have become better at understanding and forecasting volcanic behavior. They can often calculate that an eruption is likely to happen, but it is still difficult to predict the exact time and magnitude of an eruption. Scientists are extremely cautious when making predictions. They want to avoid forecasts that could cause people to evacuate and businesses to be disrupted unnecessarily, creating enormous costs to governments and society. Repeated false alarms could cause people to dismiss warnings when it's truly necessary to act, putting lives at risk.

Scientific Tools

Volcanologists use **seismographs** to monitor volcanic activity. Seismographs measure earth tremors, gas emissions, and shifts in the ground's **magnetic fields**. Seismographs record the length and strength of earth tremors, which often precede an eruption. **Satellite** technology allows scientists to measure gas emissions escaping from volcanic **vents**. **Tiltmeters** measure changes in the shape of a volcano's slopes caused by changes in gas or magma pressures. Information gathered from these help scientists create more accurate forecasts of possible volcanic eruptions.

Protecting People

Since it is impossible to stop the potentially life-threatening risks of volcanoes, our best bet is to understand the risks and to be prepared.

Observatories regularly monitor volcanic activity and issue public warnings if necessary.

Role of Observatories

Many countries with active volcano sites have built **observatories** to monitor and assess volcanic activity and the dangers it poses. The World Organization of Volcanic Observatories (WOVO) coordinates the activities of these national organizations. It provides assistance and guidance with research, monitoring, and the issuing of public warnings in the event of a possible eruption.

Role of Local Governments

Local governments in volcanic regions are responsible for developing, communicating, and carrying out emergency plans to help protect their populations. They provide education and information to their citizens about the dangers of nearby volcanoes so that people can be prepared in the event of a volcanic emergency. If a volcano erupts, they advise local populations about the best evacuation routes, road closures, dangers from avalanches or mudslides, and other potential hazards.

Be Prepared

For people living in volcano danger zones, there are some simple steps to help keep them safe. They should know where and how to obtain public safety information to help make decisions about what actions are necessary, particularly in the event of an evacuation. People should keep an emergency kit on hand. It should include goggles, breathing masks, a flashlight, and bottled water. Staying indoors in order to avoid falling ash is advised. If people do have to go outdoors, they should wear long-sleeved tops and long pants to protect their skin.

PROTECTING POTENTIAL VICTIMS

The U.S. Geological Survey (USGS) is an organization that uses an alert-level system to classify U.S. volcanoes in one of three levels: "quiet," "unrest," or "eruption." Public notifications are issued through five regional observatories to alert the public to any potential dangers of eruption.

As the human population expands, more and more people are living near active volcanoes such as Popocatépetl, near Mexico City.

LUIS'S STORY: Starting Over

As the weeks passed, help arrived to begin the long work of repairing and rebuilding. International organizations provided assistance and money to relocate survivors. Our lives changed forever. Over time, we learned that the rest of our family had died during that terrible night. Alessandro and Mateo and their families survived, but they left to live with other relatives many hours away. I often have nightmares about that dreadful night. I am sure I will never forget it.

We lost so much to the volcano. Armero lost so much. My father could no longer work because more than half of the farms and farm animals were destroyed in the mudslide. I had no school—nobody had a school. Papa said that 50 schools and two hospitals were totally destroyed. We also had no home. Ours was just one of over 5,000 buried beneath the mud. We had to start our lives all over from nothing.

Our lovely town of Armero was completely destroyed and was never rebuilt. Like other survivors, our family was relocated to a town called Lerida, 9 miles (15 km) away. The government helped us build a new home, and Papa found work on a coffee plantation. Nicolas, Maria, and I go to a different school where we have had to make new friends. It has not been easy.

Coffee is an important crop in Colombia and is grown in the region around Nevado del Ruiz. Many of the workers had to find new jobs when the farms were destroyed.

After the tragedy, many people were angry that the government and town officials had not warned us about the dangers of Nevado del Ruiz, the Sleeping Lion. They wondered how many people could have been saved if the volcano had been watched more closely and authorities had sounded the alarm sooner. Many experts agreed that this was a tragedy that could have been avoided. Knowing this has made us incredibly sad.

The town of Armero was left in ruins after the eruption.

Sometimes we go back to Armero. Where the town once stood, there is now a memorial center for the 20,000 people who were killed that day. I know we are among the lucky ones—the survivors. We will never take the Sleeping Lion for granted again.

Glossary

acid rain Rainfall made so acidic by atmospheric pollution that it damages the environment

atmosphere The layer of gases that surrounds a planet

avalanches Masses of snow, ice, and rocks that slide suddenly down a mountainside

cinder A small piece of burned material

crater The bowl-shaped hole at the top of a volcano

debris Waste or pieces of material left over from an event such as a disaster

deposits Things placed or left behind

erupts Explodes

fertile Land that is suitable for growing crops

fictional Made up, not true

gravitational Relating to the force of gravity, which pulls objects toward Earth

hot spot The area where magma streams up under the surface from deep down

lahar A violent type of mudflow made from volcanic material, rocky debris, and water

landslides The sliding of loose earth and rock down a steep slope

lava Melted rock from a volcano

magma Hot, liquid rock

magnetic fields Part of the forces that control Earth's gravity

mudslides Masses of mud and other earthy material that fall down a hillside or other slope

natural disaster A disaster caused by nature, not human-made

observatories Special buildings that have large telescopes for studying the weather, volcanoes, or the solar system

orbit The regular path of a celestial body or spacecraft around a star or planet

plume Something that rises into the air in a tall, thin shape. It can also be the column of rising hot magma in a mantle that creates a hot spot.

satellite A human-made object that sends information from space to Earth

seismographs Instruments that measure movement beneath Earth's crust

tectonic plates The pieces of Earth's crust that fit together

tephra Rock fragments and particles ejected in a volcanic eruption

tides The movement of the ocean that is caused by the pull of the Sun and the Moon on Earth

tiltmeters Instruments that measure changes in the level of Earth's surface

toxic Poisonous

tremors Small earthquakes

tsunamis A series of large waves caused by an underwater earthquake

vents Openings through which air or steam is released

volcanic bombs Any chunks of molten rock larger than 2.5 inches (6.4 cm) that are spewed out of a volcano

Learning More

Find out more about volcanoes and their dangers.

Books

Howell, Izzi. *Volcano Geo Facts.* Crabtree Publishing, 2018.

Hudak, Heather C. *Volcanic Eruption Readiness.* Crabtree Publishing, 2020.

Nargi, Lela. *Absolute Expert: Volcanoes.* National Geographic Kids, 2018.

Tarshis, Lauren. *I Survived the Eruption of Mount St. Helens, 1980.* Scholastic, 2016.

Websites

Watch great volcano videos and find out more about volcanoes at:
https://easyscienceforkids.com/all-about-volcanoes

Explore explosive volcano facts at:
https://kids.nationalgeographic.com/explore/science/volcano/

Discover the latest updates on active volcanoes, detailed information on how volcanoes work, and fun facts about volcanic eruptions at:
https://volcano.si.edu/#

Learn about some famous volcanic eruptions at:
www.weatherwizkids.com/weather-volcano.htm

Index

About the Author

Linda Barghoorn has traveled to more than 60 countries, shared experiences with extraordinary people, and captured memories through all the photos she's taken along the way. She believes that learning about people, places, and things can teach us about ourselves and how we can change the world.